NATURE IN FOCUS

LIFE ON THE TUNDRA

By Jen Green

Gareth Stevens
Publishing

Please visit our Web site www.garethstevens.com. For a free color catalog of all our high-quality books, call toll free 1-800-542-2595 or fax 1-877-542-2596.

Library of Congress Cataloging-in-Publication Data
Green, Jen.
  Life on the tundra / Jen Green.
    p. cm. — (Nature in focus)
  Includes index.
  ISBN 978-1-4339-3417-9 (library binding) — ISBN 978-1-4339-3418-6 (pbk.)
  ISBN 978-1-4339-3419-3 (6-pack)
  1. Tundras—Juvenile literature. 2. Tundra ecology—Juvenile literature. I. Title.
  GB571.G74 2010
  577.5'86—dc22   2009037144

Published in 2010 by
Gareth Stevens Publishing
111 East 14th Street, Suite 349
New York, NY 10003

For Gareth Stevens Publishing:
Art Direction: Haley Harasymiw
Editorial Direction: Kerri O'Donnell

For The Brown Reference Group Ltd:
Editorial Director: Lindsey Lowe
Managing Editor: Tim Harris
Editor: Jolyon Goddard
Children's Publisher: Anne O'Daly
Design Manager: David Poole
Designer: Lorna Phillips
Picture Manager: Sophie Mortimer
Picture Researcher: Clare Newman
Production Director: Alastair Gourlay

Picture Credits:
Front Cover: Shutterstock: George Burba (main image); TT photo (background)
FLPA: Michio Hoshino/Minden Pictures: 23, 25; David Hosking: 8; Minden Pictures: 20; Mark Newman: 17; istockphoto: Cushkin: 14; Shantell: 31; Siverwebs Studio: 11; Walter Spina: 29b; Jupiter Images: Photos.com: 3, 18, 21; Stockxpert: 13t; Shutterstock: 13b; John A. Anderson: 15; Roger Asbury: 12; Vesley Aston: 24; Edward Bruns: 10-11; Magdalena Bujak: 19t; George Burba: 5; Vladimir Chemyanskiy: 19b; Dennis Donohue: 4; John M. Fugett: 16; Kirk Geisler: 26; Kaido Karner: 27: Dean Mitchell: 30; Maxim Petrichuk: 29t; TT Photo: 9
All Artworks Brown Reference Group

Manufactured in the United States of America
1 2 3 4 5 6 7 8 9 12 11 10

CPSIA compliance information: Batch #BRW0102GS: For further information contact Gareth Stevens, New York, New York at 1-800-542-2595.

Contents

Tundra Around the World

The tundra is a belt of treeless lowlands that is **circumpolar** because it follows the **Arctic Circle** and extends around the entire globe. The Arctic Ocean lies north of the tundra and is covered by ice. The taiga, a belt of **coniferous** trees, lies south. This book introduces the plants and animals that live on and around the tundra.

A lynx hunts on the tundra. Its pale spotted winter coat blends in with the snowy landscape.

The tundra stretches across North America, Europe, and Siberia, in and around the Arctic Circle.

In summer, the snow and ice melt. Meltwater collects on the tundra, forming marshes, lakes, and pools.

Life on the Tundra

polar bear

musk ox

snowy owl

wolf

Animals that live in the southern part of the tundra all year include mammals such as the Arctic fox and birds such as the ptarmigan and gyrfalcon.

peregrine falcon

ptarmigan

Arctic hare

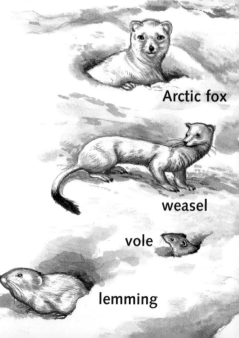

Arctic fox

weasel

vole

lemming

Some animals live on the tundra all year round. Others stay for summer and **migrate** south for winter. As Earth moves around the Sun, it tilts on an **axis**. When the North Pole leans toward the Sun, it is summer on the tundra, and there is almost 24 hours of daylight. In winter, when the North Pole tilts away from the Sun, the tundra is dark.

cranes

Arctic tern

Animals such as geese, grizzly bears, moose, and some caribou can survive on the tundra only in summer.

grizzly bear

moose

lynx

Tundra plants are tough enough to survive freezing winters. They grow close to the ground, protected from icy winds.

wolverine

caribou

Arctic willow

Arctic lupin

snow goose

curlews

Arctic poppy

mosquito

lichens

midges and gnats

Plants and Insects

Hundreds of different plants and insects live on the tundra. They are tough enough to withstand the long, dark months and bitter cold of an Arctic winter. Only dwarf trees survive on the open tundra. Instead of growing upward, these tiny trees grow more like shrubs, spreading their branches along the ground, out of the wind.

Dwarf trees, like this Arctic willow, have big flowers to attract insects.

When summer comes, tundra plants grow and flower quickly, and the sound of buzzing insects fills the air.

No tall trees grow on the tundra. The dark spruce and fir trees thin out at the edge of the taiga.

Saxifrage is a small, low-growing herb. It starts growing in early spring and flowers even before the snow has melted.

Dwarf willow trees measure only 4 inches (10 cm) tall, but their branches can stretch as far as 16 feet (5 m) along the ground.

Keeping Warm

Plants, such as saxifrage and moss campion, grow in low, dense groups. In winter, these herbs escape the killing frosts under a protective blanket of snow. Like many other flowering plants that bloom on the tundra, Arctic poppies cover south-facing slopes or grow in sheltered hollows, away from the howling winds.

Collecting Light and Water

Like almost all plants, tundra plants use their roots to anchor themselves in the soil and to get moisture and minerals. The roots of tundra plants cannot grow through the layer of ground that is frozen all year—called **permafrost**—beneath the soil. So they have shallow roots that fan out in the thin topsoil. Plants make their own food using sunlight in the process called **photosynthesis**. Small, dark leaves of tundra plants absorb sunlight quickly. Some plants angle their leaves to catch the low, slanting rays of the Sun.

Arctic lupins are perennial plants that grow and bloom quickly every summer.

The Growing Season

Tundra plants have a short growing season. In the far north, this season lasts for about eight weeks in July and August, when daytime temperatures rise above freezing. Most Arctic plants are **perennial**—they live for many years. Others, such as poppies, are **annual**—they grow, flower, produce seeds, and die in one season.

Mosses and Lichens

In the far north of the tundra, only mosses, lichens, and some grasses can survive the very harsh conditions. Mosses are simple plants that grow in dense, soggy mats that cover the ground. Lichens may look a little like mosses but are in fact made up of two different types of living things—a **fungus** and an **alga**. Each depends on the other for survival. The fungus provides shelter and moisture for the alga, which like a plant can use photosynthesis to make food.

In summer, the tundra is a blaze of colors, with pink, purple, and yellow flowers carpeting the ground.

Gray lichen grows among Arctic bearberry on the Alaskan tundra.

WET AND DRY

Drought is a problem on some areas of the tundra. The air is cold and dry, and little rain falls. For these reasons, the tundra is also called a polar desert. Tundra plants in these regions suck every drop of precious moisture from the bone-dry ground. Other areas of the tundra are swampy, because water cannot drain through the frozen permafrost. Marsh plants such as mosses, heathers, and Arctic cranberries (shown left) flourish there.

Reindeer Moss

Some lichens grow on stones and boulders. Other lichens grow in thick mats. One of these, called reindeer moss, is a yellow lichen that looks like thousands of tiny reindeer—another name for caribou—antlers. It provides food for caribou in winter when other food is scarce.

Many butterflies live on the southern part of the tundra. Because summer is so short, their life cycles take years to complete.

Some crawling insects, spiders, and mites can survive anywhere on the tundra.

Tundra Animals

Insects and other small animals, such as spiders, also live on the tundra. Arctic insects include flies, beetles, moths, and butterflies. Unlike warm-blooded birds and mammals, these animals are cold-blooded. This means that to keep warm, they have to take heat from their surroundings, and to keep cool, they lose heat to their surroundings. In freezing weather, small animals from warmer lands cannot stay active, yet tundra **species** can.

Pesky Insects

In summer, large clouds of mosquitoes descend on caribou and other warm-blooded animals to drink their blood. Their large prey provide food and warmth for the insects. Botflies lay their eggs inside the noses of caribou. When the botfly maggots hatch, they burrow up into the deer's nose to feed and spend winter in warmth and safety.

Mosquitoes, midges, blackflies, and gnats are common around tundra pools in summer.

ARCTIC SURVIVAL

The blood of some tundra insects contains a substance that keeps them from freezing. Many other insects on the tundra have dark-colored bodies that absorb heat from the Sun more quickly than pale colors do.

The life cycles of tundra insects coincide with the seasons. These insects hatch, grow, and breed in the short period of spring and summer. They lay tough-shelled eggs that survive winter and hatch the following spring.

Year-Round Residents

Some birds and mammals can live on the tundra all year. Birds and mammals are warm-blooded animals. Food provides the fuel that keeps their bodies warm in freezing weather. Year-round residents include snowy owls, gyrfalcons, Arctic foxes, Peary caribou, musk oxen, lemmings, weasels, and polar bears.

On the southern tundra, a ground squirrel emerges from a deep winter sleep called hibernation.

Musk oxen are huge animals. A musk ox can weigh up to 880 pounds (400 kg) and stand up to 5 feet (1.5 m) tall.

Musk oxen eat grasses, mosses, and lichens. They dig out food from under the snow.

Feathers and Fur

In fall, mammals and birds eat plenty of food and build up a layer of body fat that will keep them warm in winter. Birds grow a dense coat of feathers with two layers—a tough, waterproof layer of outer feathers, and a soft, warm, downy underlayer. Mammals grow a thick coat of fur with shaggy outer hairs and woolly underfur. Musk oxen have the longest fur of any mammal—up to 3 feet (1 m) long. Caribou, Arctic foxes, and polar bears all have hollow hairs that trap warm air.

Rodents, such as collared lemmings, live in snowy burrows during winter. The temperature there is much warmer than aboveground.

Keeping the Heat In

Arctic mammals, such as hares and foxes, have compact bodies. With their short ears and legs, they lose less heat than long-eared, long-limbed animals from warmer lands. Arctic foxes have hairy toes that keep them warm as they pad across the snowy landscape.

In their white winter coats, weasels are called ermines. Ermines are sometimes trapped for their fur.

Winter Colors

In winter, the snowy landscape offers little cover for animals to hide from prey or predators. Smaller birds and mammals grow a white winter coat of fur or feathers. This covering blends in with their surroundings, making them difficult to see. Ptarmigan have mottled brown feathers in summer but white winter plumage. Arctic foxes, hares, and weasels have brown or gray fur in summer but a pure white winter coat.

Tundra Prey

Rodents such as voles and lemmings are common on the tundra. Plant buds, shoots, roots, and other plant matter are popular food choices for these small animals. In summer, when plants are abundant, lemmings breed quickly, and their populations expand rapidly. Their crowded communities are suddenly revealed when the snow melts the following spring, making the lemmings easy prey for owls, falcons, and wolves.

The willow ptarmigan is a ground-living bird that spends the whole year on the tundra.

The Arctic fox has very good hearing. This one digs out a rodent that it can hear moving under the snow.

The polar bear is a big hunter—up to 8 feet (2.5 m) long. Despite its size, it can run fast in short bursts when hunting.

Tundra Hunters

Wolves are among the top hunters of the tundra. These wily animals usually hunt in packs of eight to twenty. On a hunt, wolves creep up on caribou and pick out the young, isolated, or sick as their prey. They surround the herd and lunge forward to separate their prey from the rest of the herd. Musk oxen protect their young from wolves by forming a circle with the calves on the inside. Polar bears sometimes roam the tundra, although they generally hunt on the coasts and far out on the sea ice. They are expert swimmers. The bears kill seals, fish, and land mammals with their sharp teeth and powerful paws.

Working together in packs, wolves are able to kill prey as large as a musk ox.

CHAIN OF LIFE

Tundra plants and animals depend on one another for food. The relationship between these animals is called a food chain (shown right). All tundra plants and animals are part of the tundra **ecosystem**. Plants and lichens make their own food through photosynthesis. Herbivores such as lemmings, ptarmigan, and caribou feed on plant matter. Some plant eaters are hunted by smaller **carnivores**, such as weasels and falcons. These in turn fall prey to the top predators of the tundra—wolves and polar bears. When animals die, their bodies are eaten and broken down by insects, fungi, and tiny bacteria. This process is called decomposition. Animal remains enrich the soil to help plants grow, and so the cycle comes around again.

lemming

weasel

wolves

A snowy owl brings food to its chicks. Unlike many other owls, snowy owls hunt during the day.

Tundra Birds

Few birds are hardy enough to last all winter on the tundra. Ptarmigan scratch for seeds and buds under the snow with their beaks and feet. Ravens scavenge for scraps of meat in winter. In summer, they raid other birds' nests to steal their eggs and chicks.

Gyrfalcons and snowy owls are year-round predators in parts of the tundra where there is enough sunlight for hunting. They swoop low and seize small mammals with their sharp, hooked claws. In spring, snowy owls raise their chicks on a diet of lemmings.

In winter, small birds called redpolls and buntings search for seeds on the southern tundra.

Summer Visitors

Migratory animals arrive on the tundra in spring and take advantage of the summer climate and plentiful food. From the south, they travel many miles to breed in this remote place, far from people. On the treeless plains, animals are safer from the many predators because there is more food available to the predators.

After their winter sleep, grizzly bears search for food on the tundra.

Caribou herds are made up of females and their young, led by the older, experienced females.

Great herds of wandering caribou roam the treeless plains of the tundra during summer.

Millions of Migrants

Summer visitors include mammals such as caribou, grizzly bears, wolverines, red foxes, and lynx. Migrating birds include ducks, geese, swans, and cranes, waders such as curlews and phalaropes, snow buntings, snowy owls, and falcons. Migrating animals travel in herds or flocks, led by experienced adults. Young animals follow their parents, who use familiar landmarks to find the way. In fall, both adults and their offspring begin the long trip south again.

Wolverines are large, fierce weasels. In summer, they hunt prey—often much bigger than themselves—on the tundra.

Moose—the world's largest deer—visit the tundra in summer and graze on water herbs.

Feathered Visitors

More than 100 species of birds visit the tundra in spring and summer. Some arrive in April, as the snow melts, to nest and lay their eggs on the marshes. Many have already mated before they arrive. The others attract their partners with courtship dances and calls.

The tundra becomes busy in spring, with the whoops, whistles, and quacks of nesting birds. Different kinds of birds can nest close together because they eat different foods and do not need to fight for precious morsels. Wading birds feed on insects, while ducks and geese nibble water plants. The chicks hatch in midsummer, when food is most abundant.

Cranes fly in to spend summer on the tundra. They hunt fish and worms in pools and streams.

The Return South

Chicks have to grow fast because they will soon need to fly south. The parents lead their young to food or spend the long summer days finding morsels for them. The young birds practice their swimming and flying skills for the long trip south.

Golden eagles hunt by swooping low over the treeless plains of the tundra in summer.

GRACEFUL FLIERS

Snow geese are common tundra nesters. They are easy to identify, with their snow-white bodies and black wing tips. From their winter homes in the Gulf of Mexico, the geese fly 2,000 miles (3,200 km) north to breed in the Arctic. They can be seen flying overhead in V-shaped groups when traveling to and from the tundra. The geese take turns at the tip of the V, where it takes the most energy to fly.

Golden eagles hunt hares, grouse, and lemmings.

Marshland Birds

The coastal marshes are home to birds such as loons and eider ducks. During the long daylight hours of summer, these birds bob and dive in the choppy waters. As fall approaches the days become shorter. Then, the flocks fly away and the marshes are once again deserted.

A golden plover visits a coastal region of the tundra in summer.

When a hunt has been successful, you may see a golden eagle soaring upward with prey grasped in its claws.

The Tundra on Your Doorstep

The tundra may be far away, but birds near your home may spend summer on the tundra and winter in your area—or pass by on their way south. You can look in books and on websites for more details about such birds. Watch for geese, swans, ducks, and other migrants overhead, and study them with binoculars.

Squirrels act like tundra animals—gathering nuts in fall to get them through winter.

Animal Tracks

In winter, **biologists** study tundra animals by looking at their tracks in the snow. Animal tracks can also be found in the mud by ponds and streams at any time of year. A wildlife book will help you identify the animals that made the imprints. Study the tracks to see how the animals were moving—were they walking or running?

Like tundra wildlife, animals near you behave differently according to the seasons. In fall, they prepare for winter. You might see small mammals, such as squirrels, burying nuts.

TOP TIPS FOR BIRD WATCHERS

1 Dress in warm, waterproof clothing when you go bird watching or tracking. Always take an adult with you to keep you safe.

2 Stand still when you watch migrating birds with binoculars so you do not trip and hurt yourself.

3 Dull-colored clothing helps you get close to birds on the ground. Avoid sudden movements that might scare the animals away.

If you live near the routes of migrating birds, you may be able to see them on their journey.

Glossary

alga: a tiny plantlike form of life

annual: a plant that completes its life cycle in a single season

Arctic Circle: an imaginary line encircling Earth a few hundred miles south of the North Pole

axis: an imaginary line that passes through Earth from the North Pole to the South Pole

biologists: scientists who study living things such as animals

carnivores: animals that eat flesh

circumpolar: surrounding or located in one of the polar regions

coniferous: a word describing needleleaved trees such as pines and spruces

drought: a long period of time with not enough rain

ecosystem: a community of living things

fungus: a type of living thing that belongs to a group including toadstools and mushrooms

migrate: to make a regular journey. Animals such as birds or moose migrate to escape bad weather, find food, or breed

perennial: a plant that continues growing for a number of years

permafrost: the permanently frozen ground that lies beneath the topsoil in very cold places

photosynthesis: how plants and algae make food using sunlight

species: a group of closely related animals, plants, or other living things

Index